J. WESTON
# WALCH
PUBLISHER
Portland, Maine

# Content-Area Vocabulary Strategies

## Mathematics

Josh Brackett

# User's Guide
## to
## *Walch Reproducible Books*

Purchasers of this book are granted the right to reproduce all pages where this symbol appears.

This permission is limited to a single teacher, for classroom use only.

Any questions regarding this policy or requests to purchase further reproduction rights should be addressed to:

Permissions Editor
J. Weston Walch, Publisher
321 Valley Street • P.O. Box 658
Portland, Maine 04104-0658

1    2    3    4    5    6    7    8    9    10
ISBN 0-8251-4339-X

# Contents

# INTRODUCTION

*Content-Area Vocabulary Strategies* gives students tools to decipher unfamiliar words in their content-area reading—or any reading.

Experienced readers use vocabulary strategies without thinking about them—they have become second nature. But for students just making the transition from fiction to nonfiction, from reading for pleasure to reading to learn, these skills are new. This book breaks down the deciphering process into manageable strategies that will help students read in any content area. Having concrete strategies to turn to gives students the confidence to succeed in understanding whatever they are asked to read.

The target vocabulary words in the lessons have been compiled from various lists and textbooks to offer a strong base of common and important mathematics words. Each lesson is self-contained, with a reading showing the vocabulary words in context, a context-clue activity, an extension activity, and a writing activity.

## Classroom Management

*Content-Area Vocabulary Strategies* is easy to use. Simply photocopy each self-contained lesson, consisting of four parts, and distribute it. The Answer Key at the back of the book makes scoring simple and quick.

First photocopy and distribute the introductory lesson, "Context Clues." This names the types of clues and shows them in action. This is a model that students can emulate in their own thinking and reading.

Then photocopy each lesson and work through the lessons. Each lesson has four parts:

**Activity 1: Introducing Vocabulary in Context**—A short, informative article shows the vocabulary words in context. Students are asked to find the context clues that reveal the meaning of the vocabulary word.

**Activity 2: Developing Vocabulary in Context**—Students practice using context skills to discover word meanings.

**Activity 3: Extending Vocabulary Strategies**—Games, puzzles, and other activities practice and extend the lesson vocabulary.

**Activity 4: Writing to Build Vocabulary**—A writing topic related to the reading provides a forum for incorporating new vocabulary in students' writing. Writing with the new words helps anchor them in the students' knowledge base.

The Word Journal suggested at the end of each Activity 4: Writing to Build Vocabulary is a great way for students to take ownership of new and interesting words. This is optional. You may wish to compile a class word journal, accepting suggestions from everyone and keeping an ongoing list throughout the school year.

# INTRODUCTORY LESSON
## Context Clues

Reading, writing, speaking, and listening are complex processes. Whenever you do one of these tasks, you use a variety of skills to understand or create meaning. One of the most important parts of understanding language is figuring out vocabulary.

Vocabulary refers to the words you encounter throughout your life. *Your* vocabulary is every word you know. The English language is so rich that you will probably never stop gaining new vocabulary. Besides words already in existence, new words are continually coined. And old words take on new or different meanings. Learning to understand unfamiliar vocabulary will serve you well throughout your life.

One way to learn new vocabulary is to look it up in a dictionary. Or you can ask someone else what a word means. A more efficient way—because it does not require you to interrupt your reading—is to look for context clues.

### Context Clues

Context refers to all the words around an unfamiliar word. Context also includes your personal experience and knowledge. Clues are hints or pointers that help you uncover unknown information. Context clues, then, are hints about the meaning of a new word that you find in the text around that word.

Here are some types of context clues.

**Restatement.** A word may be clarified by saying it another way—in other words. The restatement may appear in the form of a synonym—a word that means almost the same thing. The restatement may also be a phrase.

Example: The <u>sum</u>, or total, of the bank deposits was $580.

**Antonym.** Just as a synonym can clarify an unfamiliar word by comparing it to something you know, an antonym can help you figure out a new word by contrasting with something you recognize.

Example: Unlike addition, <u>subtraction</u> reduces the number of items in a set.

# Context Clues *(continued)*

**Definition.** A word may be followed by a straightforward definition, as you would find in a dictionary.

>Example: The <u>sum</u> (the result of adding together) of the bank deposits was $580.

**Explanation.** An explanation is similar to a definition, without stating directly what a word means.

>Example: A <u>sum</u> results when you add numbers together.

**Example.** Another way to communicate the meaning of a word is by giving an example or examples. Some words that can signal that an example is being given are include/including, such as, like, and for example.

>Example: The math problem required many <u>operations</u>, including addition, subtraction, and multiplication.

**Description.** A description tells what something is like, without giving a direct definition.

>Example: The club's monthly <u>balance</u> increased as new members joined and paid dues. Not only did the number of dues grow larger; the cost of dues was increased at the beginning of the year. This means that the balance was even greater than had been expected, because the old dues were in effect at the time of the financial forecast.

Read the following passage. As you read, try to figure out the meaning of the underlined words, using context clues. Then answer the questions that appear after the passage.

Creating and sticking to a <u>budget</u> can help you make wise decisions about your money. A budget is a spending plan. It helps you balance your income and expenses.

To make a monthly budget, first <u>tally</u>—add up—all your income. <u>Income</u> includes all the money you take in, such as a regular paycheck and an allowance.

Once you have a figure for your income, you need to add up your <u>expenses</u>. Regular expenses such as a monthly haircut or weekly DVD rental can be <u>calculated</u> pretty accurately.

Unless your stylist or video store increases prices <u>dramatically</u>, you will still have a good number to work with. <u>Incidentals</u>, such as a special birthday gift for a best friend, or a meal out to celebrate, are not <u>fixed</u>. Expenses that vary are difficult to plan for. <u>Deduct</u>, or subtract, the regular expenses from your income, and you will know how much money you have left per month to <u>splurge</u> on incidentals. If you get a negative number, then you have a problem. You will either have to increase your income, or decrease your expenses.

# Context Clues (continued)

1. What is a budget?

   What context clues helped you figure out the meaning of the word?

2. Define *tally*.

   What kind of context clue gave you the meaning?

3. What is income?

   What examples are given of income?

4. What are expenses?

   What type of context clue helped you understand what expenses are?

5. What does *calculate* mean?

   What context clues helped you figure out the meaning?

6. Write a definition or a synonym for *dramatically*.

   What context clues helped with your definition?

7. What does *incidentals* mean?

   What context clues gave you hints about the meaning?

8. Define *fixed*.

   How did you figure out the meaning?

9. What does *deduct* mean?

   What type of context clue gave you the meaning?

10. Define *splurge*.

    How did you figure out the meaning?

# LESSON 1
## Counting in an On/Off Universe

**Activity 1** **Introducing Vocabulary in Context**

Read the following article. Notice the words in bold type. Use context clues to figure out the meaning of these words.

The *I Ching* or "Book of Changes" dates from sometime between 1000 and 500 B.C. and is based on a Chinese tradition that is said to be as old as 10,000 years. The *I Ching* was used by the ancients to foretell the future and is still in use today. It provides a set of 64 *kua* or hexagrams. Each hexagram is made up of six horizontal lines. Each line is either straight (*yang*) — or broken (*yin*) – –.

For example, in this hexagram, which is called *Hsiao Ch'u* or "The Taming Power of the Small," the fourth line up from the bottom is broken; all the others are straight.

To tell your fortune, you choose one of the hexagrams by some **random** process like tossing coins. Then you look in the *I Ching* for the hexagram's meaning and try to apply it to your own life.

Why are there 64 hexagrams? Each hexagram has six lines. Each line has two possibilities: either broken or straight, yin or yang. So there are $2^6$ or 64 possible hexagrams.

Why are there only two possibilities for each line? The ancient Chinese seem to have understood that we live in a two-state or on/off universe. The first game that every child learns is peek-a-boo: now you see me; now you don't. Many things in our world alternate between two states: day/night, school/vacation, waking/sleeping. Objects are surrounded by space (thing/nothing). At the subatomic level, the world is made of particles and the space between them (particles/no particles) or of waves moving up and down.

In electrical circuits, current is either flowing or not flowing: on/off. That's why computers, which run on electricity, store numbers using the **binary system.** To understand what that means we first need to look at the numerals that we usually use, which are based on the decimal system.

The **decimal system** originated in India. It was brought to the West by the Arabs. This is why the numerals we use are called Arabic. The **base** of the decimal system is ten. Why ten instead of some other number? Undoubtedly this is because the earliest counting device was the ten fingers that are standard equipment for the human body.

*(continued)*

# Counting in an On/Off Universe *(continued)*

The decimal system has ten digits that stand for the first ten whole numbers: 0, 1, 2, 3, 4, 5, 6, 7, 8, and 9. (In fact, the word **digit** comes from the Latin word for finger.) To represent numbers greater than 9, we use **place value,** which simply means that the place where we put a digit tells us what its value is.

For example, the numeral 283 in **expanded form** is $200 + 80 + 3$, which equals $2{\times}100 + 8{\times}10 + 3{\times}1$. We know that the 3 means $3{\times}1$ or 3 because it is in the rightmost place. The next place to the left is the tens place. Whatever digit is there is to be multiplied by ten. The next place to the left is the hundreds place, and so on. The value of each place is ten times that of the place to the right of it: the next higher **power** of ten. 283 means $2{\times}10^2 + 8{\times}10^1 + 3{\times}10^0$. (Any number with a zero **exponent** equals 1.) 9,437, 236 means $9{\times}10^6 + 4{\times}10^5 + 3{\times}10^4 + 7{\times}10^3 + 2{\times}10^2 + 3{\times}10^1 + 6{\times}10^0$.

The binary system has two digits that stand for the first two whole numbers: 0 and 1. (In a computer, 0 and 1 are off and on.) Like the decimal system, the binary system uses place value to represent numbers greater than one. The first place on the right is the ones place. Whatever digit is there is to be multiplied by one, which means it stands for itself. The next place to the left is the twos place. Whatever digit is there is to be multiplied by two. The next place to the left is the fours place, and so on. The value of each place is two times that of the place to the right of it: the next higher power of two.

So the number that we would write as 283 in the decimal system we would write in the binary system as $100011011_2$. (The subscript $_2$ means that it's a base two numeral, not base ten.) $100011011_2$ means $1{\times}2^8 + 0{\times}2^7 + 0{\times}2^6 + 0{\times}2^5 + 1{\times}2^4 + 1{\times}2^3 + 0{\times}2^2 + 1{\times}2^1 + 1{\times}2^0$. If we didn't know what $100011011_2$ meant, we could **evaluate** it in this way:

| | | | | |
|---|---|---|---|---|
| $1{\times}2^8$ | = | $1{\times}256$ | = | 256 |
| $0{\times}2^7$ | = | $0{\times}128$ | = | 0 |
| $0{\times}2^6$ | = | $0{\times}64$ | = | 0 |
| $0{\times}2^5$ | = | $0{\times}32$ | = | 0 |
| $1{\times}2^4$ | = | $1{\times}16$ | = | 16 |
| $1{\times}2^3$ | = | $1{\times}8$ | = | 8 |
| $0{\times}2^2$ | = | $0{\times}4$ | = | 0 |
| $1{\times}2^1$ | = | $1{\times}2$ | = | 2 |
| $1{\times}2^0$ | = | $1{\times}1$ | = | 1 |
| | | | | 283 |

# Counting in an On/Off Universe *(continued)*

**Activity 2**    **Developing Vocabulary in Context**

Read each sentence below. Choose the word or phrase from the box that best completes each sentence.

| | | |
|---|---|---|
| random | digit | power |
| binary system | place value | exponent |
| decimal system | expanded form | evaluate |
| base | | |

1. 10000000001 is a(n) _____ numeral with 11

   _____s. When I _____ it, I see that

   it equals 1025 in the _____.

2. 5000 + 400 + 30 + 8 is 5438 in _____. We know that the 4

   in 5438 stands for 400 because of its _____.

3. Drawing a number out of a hat as in a raffle is an example of a(n)

   _____ process.

4. In the expression $10^6$, 10 is the _____ and 6 is the

   _____. It means the sixth _____

   of 10.

# Counting in an On/Off Universe *(continued)*

**Activity 3**      **Extending Vocabulary Strategies**

Use a dictionary to complete this activity.

1. The root of the word *binary* is the Latin word for two. Go to the page in your dictionary where *binary* is listed. Find two other words listed nearby whose meaning includes the idea of two. Write a brief definition of each.

    a. word: _____      definition: _____

    b. word: _____      definition: _____

2. The root of the word *decimal* is the Latin word for ten. Go to the page in your dictionary where *decimal* is listed. Find two other words listed nearby whose meaning includes the idea of ten. Write a brief definition of each.

    a. word: _____      definition: _____

    b. word: _____      definition: _____

3. The root of the word *evaluate* is the Latin word for worth. The word *value* comes from the same root. Go to the page in your dictionary where *value* is listed. Find two other words listed nearby whose meaning includes the idea of value. Write a brief definition of the word *value* and the two other words.

    a. word: value      definition: _____

    b. word: _____      definition: _____

    c. word: _____      definition: _____

4. The root of the word *exponent* is the Latin word for put. In English words it appears as *pon* or *pos*. Find three words that have the same root and write brief definitions of each. You will find them in your dictionary under *pon* or *pos* with no prefix or under one of the following prefixes: *com-* (together), *de-* (from, down, away), *dis-* (away, not), *ex-* (out of, away, from), *im-* (in, into), *op-* (toward, against), *post-* (after), *pre-* (before), *pro-* (for, forward), *re-* (back, again), *sup-* (under).

    a. word: _____      definition: _____

    b. word: _____      definition: _____

    c. word: _____      definition: _____

**Activity 4**     **Writing to Build Vocabulary**

Imagine that a bill has been introduced in the House of Representatives that says that starting in one year from the date when the bill is passed, everyone doing business with the U. S. Government will be required to use the binary system of numerals rather than the decimal system. If you were a member of the House, would you vote for or against the bill? Write a one-minute speech that you would make from the floor of the House explaining your vote. Use as many of the following words in your speech as you can: *base, decimal system, evaluate, exponent, power, binary system, digit, expanded form, place value, random.*

**Word Journal**

Are there words from this lesson, or other words you have found on your own, that you'd like to know more about? In a word journal, add any interesting words that you want to "make your own." Write the word and its definition. Add examples or drawings that explain the word.

# LESSON 2
# Nonrectangular Numbers

**Activity 1**    **Introducing Vocabulary in Context**

Read the following article. Notice the words in bold type. Use context clues to help you figure out the meaning of these words.

Multiply any number by 1 and the **product** you get is the number you started with. That is the **identity property of multiplication.** Because division is the opposite of multiplication, every counting number is **divisible** by itself and by 1. However, most counting numbers are divisible by two or more counting numbers other than themselves or 1. For example, 12 is divisible by 2 because the **quotient,** 6, is a counting number. In other words, most numbers have **factors** other than themselves and 1. 12 has four: 2, 3, 4, and 6. So 12 is a **composite** number.

Think of composite numbers as rectangular numbers. If a number is composite, you can make a rectangle out of it. There are two ways to make a rect-angle out of 12 black circles.

But there are some numbers that, no matter how hard you try, you can't make a rectangle out of. They're not divisible by any counting number but themselves and 1. They have no factors except themselves and 1. They're **prime** numbers.

Even if you've never heard of prime numbers before, you can probably name a few: 2, 3, 5, 7, 11, and 13 are all prime numbers.

Suppose we wanted to find all the prime numbers less than 100. How would we do it? The hard way would be to take each of the numbers from 2 to 100 one at a time and try dividing them by other numbers. There's an easier way. It's called the **Sieve of Eratosthenes.***

Here's Eratosthenes' method of finding prime numbers. He wrote out all the numbers he was interested in. (We'll do it up to 100, but you can go as high as you want.)

| 2 | 3 | 4 | 5 | 6 | 7 | 8 | 9 | 10 | 11 | 12 | 13 | 14 | 15 | 16 | 17 | 18 | 19 | 20 | 21 |
|---|---|---|---|---|---|---|---|----|----|----|----|----|----|----|----|----|----|----|----|
| 22 | 23 | 24 | 25 | 26 | 27 | 28 | 29 | 30 | 31 | 32 | 33 | 34 | 35 | 36 | 37 | 38 | 39 | 40 | 41 |
| 42 | 43 | 44 | 45 | 46 | 47 | 48 | 49 | 50 | 51 | 52 | 53 | 54 | 55 | 56 | 57 | 58 | 59 | 60 | 61 |
| 62 | 63 | 64 | 65 | 66 | 67 | 68 | 69 | 70 | 71 | 72 | 73 | 74 | 75 | 76 | 77 | 78 | 79 | 80 | 81 |
| 82 | 83 | 84 | 85 | 86 | 87 | 88 | 89 | 90 | 91 | 92 | 93 | 94 | 95 | 96 | 97 | 98 | 99 | 100 | |

*(continued)*

# Nonrectangular Numbers *(continued)*

Then he started with the first prime number and said, "If a number is a **multiple** of 2, then it's not a prime number. So I'll erase all the multiples of 2, except 2 itself."

| 2 | 3 | 5 | 7 | 9 | 11 | 13 | 15 | 17 | 19 | 21 |
|---|---|---|---|---|----|----|----|----|----|----|
|   | 23 | 25 | 27 | 29 | 31 | 33 | 35 | 37 | 39 | 41 |
|   | 43 | 45 | 47 | 49 | 51 | 53 | 55 | 57 | 59 | 61 |
|   | 63 | 65 | 67 | 69 | 71 | 73 | 75 | 77 | 79 | 81 |
|   | 83 | 85 | 87 | 89 | 91 | 93 | 95 | 97 | 99 |   |

Then he said, "If a number is a **multiple** of 3, then it's not a prime number." So he erased all the multiples of 3 that were left, except 3 itself.

Then he said, "If a number is a multiple of 5, then it's not a prime number." So he erased all the multiples of 5 that were left, except 5 itself.

Then he erased all the multiples of 7, except 7 itself.

These are the numbers that were left.

| 2 | 3 | 5 | 7 |   | 11 | 13 |   | 17 | 19 |   |
|---|---|---|---|---|----|----|----|----|----|----|
|   | 23 |   |   | 29 | 31 |   |   | 37 |   | 41 |
|   | 43 |   | 47 |   |   | 53 |   |   | 59 | 61 |
|   |   |   | 67 |   | 71 | 73 |   |   | 79 |   |
|   | 83 |   |   | 89 | 91 |   |   | 97 |   |   |

Then he said, "That's as far as I have to go. The next prime number is 11, which is greater than 10, which is the **square root** of 100. If there is a composite number between 10 and 100 that is a multiple of a number greater than 10, then its other factor must be less than 10. But we've erased all the multiples of numbers less than 10. So all the numbers left must be prime." And they are. If you don't believe it, try dividing any one of them.

*Eratosthenes was a Greek who lived from 271 to 194 B.C. He spent most of his life in Egypt and wrote about geography, philosophy, history, astronomy, and literature as well as mathematics. A sieve is a screen that you pour a mixture through to separate out coarse particles from fine ones.

# Nonrectangular Numbers *(continued)*

**Activity 2**    **Developing Vocabulary in Context**

Read each sentence below. Choose the word or phrase from the box that best completes each sentence.

| | |
|---|---|
| product | composite |
| identity property of multiplication | prime |
| divisible | Sieve of Eratosthenes |
| quotient | multiple |
| factor | square root |

1. An easy way to find prime numbers is by using a method called the

   _____.


2. 4, 6, 8, 9, and 10 are _____ numbers.


3. The _____ of 58 × 1 is 58 because of the

   _____.


4. 71, 73, and 79 are _____ numbers.


5. 9 is _____ by 3 because when you divide 9 by 3,

   the _____ is a counting number.

   3 is a(n) _____ of 9,  and 9 is a(n)

   _____ of 3.


6. The _____ of 25 is 5.

# Nonrectangular Numbers *(continued)*

**Activity 3**    **Extending Vocabulary Strategies**

Use a dictionary to complete this activity.

The prefix and root of the word *composite* come from the Latin words for *together* and *put*. Find two other words that have the same prefix and root. Write a brief definition of each one.

1. word: _____    definition: _____

2. word: _____    definition: _____

The word *factor* has meanings outside of mathematics. Using your dictionary, write two sentences that make two of those meanings clear.

3. _____

4. _____

Find two words that have the same root as *identity*. Write a sentence for each word that makes its meaning clear.

5. _____

6. _____

Find two words that have the same root as *prime*. Write a sentence for each word that makes its meaning clear.

7. _____

8. _____

Find two words that have the same prefix and root as *product*. Write a sentence for each word that makes its meaning clear.

9. _____

10. _____

**Activity 4**     **Writing to Build Vocabulary**

Here is an example of a find-the-number problem: "Find the smallest composite number that is divisible by 3, a factor of 600, and a multiple of 10." Write three find-the-number problems of your own. Write your problems so that each one has exactly one positive answer. Use each of the following words in at least one of your problems: *composite, divisible, factor, multiple, prime, product, quotient, square root.*

**Word Journal**

Are there words from this lesson, or other words you have found on your own, that you'd like to know more about? In a word journal, add any interesting words that you want to "make your own." Write the word and its definition. Add examples or drawings that explain the word.

# LESSON 3
## Real World Math

**Activity 1**      **Introducing Vocabulary in Context**

Read the following article. Notice the words in bold type. Use context clues to help you figure out the meaning of these words.

In the real world of business, science, engineering, government, art, and music, math is used to do real work with real objects that we can see and touch. This kind of math, applied math, is used to identify things, count things, arrange things in order, and especially to **measure** things.

To measure something means to assign a number to an **attribute** of that thing. For example, to measure a truck would be to assign numbers to its length, its width, its height, its capacity, its weight, the speed of its motion, the distance it has traveled, or any one of hundreds of other attributes the truck has.

To measure things, we need to have agreed-upon units of measure. For example, if you want to know whether your truck will fit under a certain bridge, you need to measure the height of the truck and the height of the bridge. You could find a stick by the side of the road and use it as your unit of measure. You could move the stick up the side of the truck one stick length at a time, counting how many sticks high the truck is. Then you could take the stick to the side of the bridge and measure its height the same way. It might work.

But what if the bridge is hundreds of miles away? You could find out how high the bridge is by looking it up in a publication or by calling somebody up. And you could compare that measurement with the height of your truck. For that you need standard units of measure.

Most of the world uses one set of standard units called the **metric** system. The metric system is based on the **meter** as the unit of distance and the **gram** as the unit of weight. For temperature, most countries that use the metric system also use the Celsius scale, which defines 0° as the temperature at which water freezes and 100° as the temperature at which it boils. For historical reasons, the United States and other English-speaking countries still use the **customary** system based on the foot as the unit of distance and the pound as the unit of weight. For temperature, we use the Fahrenheit scale, on which water freezes at 32° and boils at 212°. However, we are slowly moving to the metric system for some things. For example, at track meets, we measure the length of a race in meters, but the length of a long jump in feet and inches. We measure the power of an engine in horsepower

*(continued)*

(1 horsepower = 550 foot pounds of work per second) and its piston displacement in liters (a liter is $\frac{1}{1000}$ of a cubic meter).

Whenever we talk about measurement, **precision** and **accuracy** are important issues. That's because whenever you measure something there is always an **error.** Because we're human, we can never find the perfectly true value of a measurement: the exact distance, or the absolutely correct weight, or the true speed. We can only come close to it. Any measurement we take is always the true value plus or minus an error. We can't ever make the error go away. We can't ever know exactly how big the error is. We can only estimate it.

Some people confuse precision and accuracy. Precision is an attribute of a measuring device. Accuracy is an attribute of a measurement. A **precise** measuring device can detect tiny little differences between one measurement and another. An **accurate** measurement is very close to the true value of the measurement; in other words, its error is small. The more precise the measuring device—provided that you use it properly—the more accurate the measurement will be and the smaller the error will be.

Suppose we need to measure the size of a picture we want to frame. We don't have a ruler, so we borrow a ruler and put it along the edge of a piece of paper. We hurriedly mark off the inches on the piece of paper with a fat crayon. Some of our crayon marks are as much as an eighth of an inch wide. We give the ruler back, and we use the piece of paper as a measuring device. How close do you think we would come to the true height and width of the picture?

Because we are using a measuring device that is not very precise, our measurement of the picture will not be very accurate. If we say that the picture is $6\frac{1}{2}$ inches wide, the true width of the picture might be as little as $6\frac{1}{4}$ inches or it might be as much as $6\frac{3}{4}$ inches. As long as we are using such a crude measuring device, we'll never know. The measurement error could be as much as $\frac{1}{4}$ inch, the difference between $6\frac{1}{2}$ inches and $6\frac{1}{4}$ inches or $6\frac{3}{4}$ inches. Our measurement is accurate to the nearest $\frac{1}{2}$ inch. The **greatest possible error** is $\frac{1}{4}$ inch, one half of $\frac{1}{2}$ inch.

If we threw away our fat-crayon-on-paper ruler and bought a good quality manufactured ruler with sixteenths of an inch clearly marked on it, we could measure the picture accurately to the nearest $\frac{1}{16}$ inch. The greatest possible error would be $\frac{1}{32}$ inch or half of a $\frac{1}{16}$ inch.

Now suppose we went to a scientific instrument store and bought a very precise Swiss caliper that can measure to the nearest $\frac{1}{10,000}$ of an inch. Then the greatest possible error would be $\frac{1}{20,000}$ of an inch. This would be more precision than we really need to frame a picture.

But suppose instead that we were molecular biochemists and we needed to measure the distance between molecules. Would the Swiss caliper be precise enough? Probably not.

When you measure something, use a measuring device that has enough precision for the application. If you measure carefully, the measurement will have the accuracy you need.

# Real World Math *(continued)*

**Activity 2**　　**Developing Vocabulary in Context**

Read each sentence below. Choose the word or phrase from the box that best completes each sentence.

| | | |
|---|---|---|
| measure | gram | accuracy, accurate |
| attribute | customary | error |
| metric | precision, precise | greatest possible error |
| meter | | |

1. A(n) _____ in the _____ system equals about a third of an ounce in the _____.

2. A yard is a little less than a(n) _____.

3. Fred wanted to know if a 12-foot by 15-foot rug would fit the floor of his room. He measured the length and width of his room by pacing it off, walking from one side of the room to the other and multiplying the number of steps by 3 feet. This way of measuring is not very _____, but Fred's measurement is probably _____ enough for this purpose.

4. Weights of chemicals measured using a platform balance are usually accurate to the nearest gram. If so, the _____ is $\pm \frac{1}{2}$ gram.

5. Height, length, and maximum speed are all _____ of a truck.

6. There is always some _____ in measurements.

7. A ruler, a scale, and a thermometer are all tools that _____ things.

# Real World Math *(continued)*

**Activity 3**     **Extending Vocabulary Strategies**

### Part I

An analogy is two pairs of words that share a similar relationship. From the list of words below, choose the word that best completes each analogy.

attribute                          precision                          meter

1. Force is to motion as _____ is to accuracy.

2. Ounce is to foot as gram is to _____.

3. John is to name as weight is to _____.

### Part II

Label each of the following examples of numbers used in the real world as
**I** = a number used only to identify something
**C** = a number that represents a count of how many
**O** = a number that says where something belongs in order
**M** = a measurement

1. ___ a bank account number

2. ___ the balance in a bank account

3. ___ the number of the inning in a baseball game

4. ___ the number on a baseball player's uniform

5. ___ the pressure in an automobile tire

6. ___ the score in a baseball game

7. ___ the time of the winner of a race

**Activity 4**    **Writing to Build Vocabulary**

Imagine that a bill has been introduced in the House of Representatives. This bill says that starting one year from the date when it is passed, everyone doing business with the U. S. Government will be required to use the metric system rather than the customary system of measurement. This bill also says that all measurement must be accurate to the nearest centimeter ($\frac{1}{100}$ of a meter) or the nearest milligram ($\frac{1}{1000}$ of a gram). If you were a member of the House, would you vote for or against the bill? Would you prefer to vote on the bill as written or would you like to see it changed before you vote? Write a one-minute speech that you would make from the floor of the House giving your point of view on the bill. Use as many of the following words in your speech as you can: *accurate, accuracy, attribute, customary, error, gram, greatest possible error, measure, meter, metric, precise, precision.*

**Word Journal**

Are there words from this lesson, or other words you have found on your own, that you'd like to know more about? In a word journal, add any interesting words that you want to "make your own." Write the word and its definition. Add examples or drawings that explain the word.

# L ESSON 4
# Visualizing Measurements

**Activity 1**     **Introducing Vocabulary in Context**

Read the following article. Notice the words in bold type. Use context clues to help you figure out the meaning of these words.

Sometimes a single measurement—for example, the distance a car has traveled—or even a series of single measurements aren't very interesting. But put one measurement together with another one—for example, how far a car can go in a given period of time—and you find out a lot more about the car.

When you put two related measurements together you get what is called an **ordered pair.** An ordered pair is simply a pair of numbers that go together in a certain order. For example, if we measured the distance a car traveled from a standing stop, we would have a series of ordered pairs of measurements: The first measurement in each pair would be the time in seconds that the car had been moving; the second, the distance it had traveled. We could write those ordered pairs in this format: (0, 0), (1, 6), (2, 13), (3, 25). Or we could write them as a table, like this:

| Time (sec-onds) | Dis-tance (feet) | Time (sec-onds) | Dis-tance (feet) | Time (sec-onds) | Dis-tance (feet) |
|---|---|---|---|---|---|
| 0 | 0 | 9 | 233 | 18 | 930 |
| 1 | 6 | 10 | 289 | 19 | 1018 |
| 2 | 13 | 11 | 352 | 20 | 1106 |
| 3 | 25 | 12 | 421 | 21 | 1194 |
| 4 | 44 | 13 | 497 | 22 | 1282 |
| 5 | 69 | 14 | 578 | 23 | 1370 |
| 6 | 101 | 15 | 666 | 24 | 1458 |
| 7 | 138 | 16 | 754 | 25 | 1546 |
| 8 | 182 | 17 | 842 | | |

Whichever way the numbers are presented, as a string of pairs of numbers or in rows and columns, it's difficult to visualize what happened with the car. A better way to see the behavior of the car is to plot the ordered pairs on a **coordinate** graph, like this:

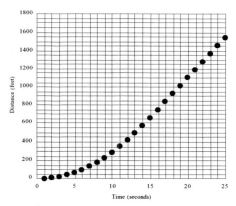

A coordinate graph is really two number lines at right angles to each other. One number line is horizontal with numbers getting bigger as you move to the right. That number line is called the *x*-**axis.** The other number line is vertical with numbers getting bigger as you move up. That number line is called the *y*-**axis.**

Every point on the graph represents an ordered pair. To find out what the ordered pair represented by a point is,

*(continued)*

draw perpendiculars from the point to the $x$-axis and to the $y$-axis. Where the perpendicular meets the $x$-axis is the first number or **$x$-coordinate** of the ordered pair (on this graph, the time in seconds). Where the perpendicular meets the $y$-axis is the second number or **$y$-coordinate** of the ordered pair (on this graph, the distance the car traveled in feet).

Plotting our ordered pairs on the graph enables us to see what the car was doing. We can see that the curve formed by the series of points starts at the **origin** of the graph—the (0, 0) point—and gets progressively steeper for the first 15 seconds. Then it straightens out. That means that for the first 15 seconds the distance traveled in each second got greater and greater. After that, the distance traveled in each second stayed the same. In other words, for the first 15 seconds of the drive, the car was accelerating. After that, its speed leveled off and it continued at a constant speed.

How fast was the car going when it got up to speed? We can tell by looking at the ordered pairs in the table starting at 15 seconds: (15, 666), (16, 754), (17, 842), (18, 930) . . . The **difference** in distance from one ordered pair to the next is 88 feet. The car was going at a constant speed of 88 feet per second.

Plotting a series of ordered pairs on a graph in this way enables us to see patterns and trends in the data that would be difficult to see by looking only at numbers. We saw that the steepness of the curve changed as the car stopped accelerating and continued at a constant speed. The steepness of the curve between one point on a graph and another is called its **slope.** The slope between one point on a graph and another is the ratio of the change in $y$ to the change in $x$.

For example, on the part of the curve where the car is accelerating, there are two points that represent the ordered pairs (2, 13) (3, 25). The change in $y$ is 12 feet; the change in $x$ is 1 second; the ratio of $y{:}x$ is 12:1. As we saw earlier, on the part of the curve where the car is up to speed, the change in $y$ for each second is 88 feet; the ratio of $y{:}x$ is 88:1.

88/1 is a positive number. The curve goes up and to the right. The slope of that part of the curve is a **positive slope.** What if the car slowed down, stopped, and backed up? As the car slowed down, the curve would bend toward the $x$-axis. When the car stopped, the curve would level off with no increase in distance from one second to the next. As the car backed up, its distance from the starting point from one second to the next would get smaller instead of bigger. The curve would go down and to the right. The change in $y$ would be a negative number. The ratio of $y{:}x$ would be negative. This is called a **negative slope.**

*Content-Area Vocabulary Strategies: Mathematics*

# Visualizing Measurements *(continued)*

**Activity 2**     **Developing Vocabulary in Context**

Read each sentence below. Choose the word or phrase from the box that best completes each sentence.

| | | |
|---|---|---|
| ordered pair | *x*-coordinate | difference |
| coordinate | *y*-coordinate | positive slope |
| *x*-axis | origin | negative slope |
| *y*-axis | | |

1. (15, 666) is an example of a(n) _____.

2. A curve or a graph that goes down and to the right has a(n)

    _____.

3. Two perpendicular number lines make up a graph. The horizontal one

    is the _____ and the vertical one is the

    _____.

4. A number that specifies the location of a point on a number line or a

    graph is a(n) _____.

5. If the ratio of the change in *y* to the change in *x* between two points on a

    graph is greater than 0, then the curve has a(n) _____.

6. In the ordered pair (6, 8), the _____ is 6 and the

    _____ is 8.

7. The _____ between 10 and 3 is 7.

8. The point on a graph whose coordinates are (0, 0) is called the

    _____.

# Visualizing Measurements *(continued)*

**Activity 3**     **Extending Vocabulary Strategies**

Read each sentence below. Choose the word or phrase from the box that best completes each sentence.

| | | |
|---|---|---|
| ordered pair | *x*-coordinate | difference |
| coordinate | *y*-coordinate | positive slope |
| *x*-axis | origin | negative slope |
| *y*-axis | | |

1. _____ is to increasing as negative slope is to decreasing.

2. Axis is to _____ as number line is to number.

3. Two related measurements together are called a(n) _____ .

4. If a curve is parallel to the _____, then *x* stays the same as *y* increases.

5. If a curve is parallel to the _____, then *y* stays the same as *x* increases.

6. In an ordered pair, the first number is the _____ and the second number is the _____.

7. Positive slope is to uphill as _____ is to downhill.

8. At a point in the Atlantic Ocean south of Ghana, the Greenwich Meridian crosses the Equator. That point is the _____ of our system of longitude and latitude.

9. Which of the following pairs of numbers is not an ordered pair? Circle the letter of your choice.

    (a) the price and quantity of something for sale, such as 6 pairs of socks for $5

    (b) the time of day, such as 2:27

    (c) the final score of a game, such as 14–7

    (d) All of the above are ordered pairs.

# Visualizing Measurements *(continued)*

**Activity 4**  **Writing to Build Vocabulary**

Using the best information available to you, construct a table of ordered pairs and draw a graph using data from your experience. Describe in words what the graph and the characteristics of the curve represent. Suggestions: your age in months and your height in inches, the price and quantity of something you would like to buy, hours you have worked and money you have earned, points scored per game. Use as many of the following words as you can in your description: *axis (axes), coordinate, negative slope, ordered pair, origin, positive slope, slope, x-axis, x-coordinate, y-axis, y-coordinate.*

**Word Journal**

Are there words from this lesson, or other words you have found on your own, that you'd like to know more about? In a word journal, add any interesting words that you want to "make your own." Write the word and its definition. Add examples or drawings that explain the word.

# QUIZ: Lessons 1–4

Circle the letter of the best answer.

1. In the expression $2^3$, the base is
   (a) 2.
   (b) 3.
   (c) 8.
   (d) 9.

2. In the expression $2^3$, the exponent is
   (a) 2.
   (b) 3.
   (c) 8.
   (d) 9.

3. $2^3$ is
   (a) eight to the second power.
   (b) eight to the third power.
   (c) three to the second power.
   (d) two to the third power.

4. Fred had only once piece of candy left. To be fair, he wanted to choose who would get it at random. Which of the following ways of choosing would *not* be random?
   (a) asking everyone to guess the number he is thinking of
   (b) drawing numbers out of a hat without looking
   (c) rolling dice
   (d) taking a vote

5. 1111 in the binary system equals _____ in the decimal system.
   (a) 13
   (b) 14
   (c) 15
   (d) 16

6. Which of the following is a composite number?
   (a) 67
   (b) 69
   (c) 71
   (d) 73

7. Which of the following is a prime number?
   (a) 36
   (b) 37
   (c) 38
   (d) 39

8. Which of the following is a factor of 14?
   (a) 2
   (b) 4
   (c) 21
   (d) 28

9. Which of the following is a multiple of 14?

   (a) 2

   (b) 4

   (c) 21

   (d) 28

10. Which of the following is the square root of 16?

    (a) 4

    (b) 8

    (c) 32

    (d) 256

11. Which of the following is *not* a unit of measure in the metric system?

    (a) gram

    (b) liter

    (c) meter

    (d) ounce

12. Sheila ran the 100-yard dash. She was clocked at 10.2 seconds. The greatest possible error of that measurement was _____ seconds.

    (a) 0.01

    (b) 0.05

    (c) 0.1

    (d) 0.5

13. That old decibel meter you are using to measure the noise level in this factory is not very _____. Your measurement will be more _____ if you use a new one.

    (a) accurate, accurate

    (b) accurate, precise

    (c) precise, accurate

    (d) precise, precise

14. Addition is to sum as subtraction is to _____ .

    (a) reduce

    (b) difference

    (c) negative slope

    (d) none of the above

15. Which of the following is not a unit in the customary system?

    (a) foot

    (b) inch

    (c) liter

    (d) pound

16. Trisha is 12 years old and has been enjoying a growth spurt. She has been measuring her height regularly and keeping records. She plotted her measurements on a coordinate graph with height in inches on the $y$-axis and the date on the $x$-axis. The slope of the curve is

    (a) positive.

    (b) negative.

    (c) neither negative nor positive.

    (d) There is no slope.

17. At the origin of a coordinate graph,
    (a) $x = 0$.
    (b) $y = 0$.
    (c) the coordinates are (0, 0).
    (d) all of the above

18. The $y$-axis of a coordinate graph is the set of all ordered pairs where
    (a) $x = 0$.
    (b) $y = 0$.
    (c) $x = 1$.
    (d) $y = 1$.

19. The difference of 3 and 5 is
    (a) 2.
    (b) –2.
    (c) 8.
    (d) 15.

20. A store tracks its total daily sales on a coordinate graph with the date on the $x$-axis and dollars on the $y$-axis. If the curve has a negative slope, it shows that business is
    (a) getting better.
    (b) getting worse.
    (c) staying the same.
    (d) can't tell

# LESSON 5
# Visualizing Data

**Activity 1**     **Introducing Vocabulary in Context**

Read the following article. Notice the words in bold type. Use context clues to figure out the meaning of these words.

The branch of mathematics called **statistics** is concerned with collecting data, analyzing it, interpreting it, and presenting it in a form that people can understand. For example, suppose there is a medical center that employs 57 people. We'd like to know how much money the people who work there earn in a year.

So we get the following data from their payroll records. The total payroll for last year was $2,922,893. The medical center employs one chief executive officer (CEO), ten physicians (MDs), 20 registered nurses (RNs), 20 technicians (Techs), and six aides. Here's the list of employees and how much they made, sorted in descending order from the highest-paid employee, the CEO, down to the lowest-paid aide.

| CEO | $254,949 | RN | $47,614 | RN | $40,544 | Tech | $27,252 |
| MD | $118,779 | RN | $47,353 | RN | $40,232 | Tech | $26,780 |
| MD | $118,553 | RN | $47,342 | RN | $40,145 | Tech | $26,157 |
| MD | $118,238 | RN | $46,814 | Tech | $29,772 | Tech | $26,095 |
| MD | $116,830 | RN | $46,711 | Tech | $29,430 | Tech | $25,862 |
| MD | $116,394 | RN | $46,104 | Tech | $29,334 | Tech | $25,738 |
| MD | $112,968 | RN | $45,325 | Tech | $28,811 | Tech | $25,689 |
| MD | $110,106 | RN | $44,682 | Tech | $28,679 | Tech | $25,501 |
| MD | $108,665 | RN | $44,368 | Tech | $28,533 | Tech | $25,433 |
| MD | $103,437 | RN | $44,072 | Tech | $28,231 | Aide | $16,816 |
| MD | $100,745 | RN | $43,062 | Tech | $28,224 | Aide | $16,598 |
| RN | $49,922 | RN | $42,580 | Tech | $28,166 | Aide | $16,446 |
| RN | $49,316 | RN | $40,916 | Tech | $27,756 | Aide | $16,166 |
| RN | $48,444 | RN | $40,845 | Tech | $27,462 | Aide | $16,029 |
| | | | | | | Aide | $15,879 |

This list is what statisticians call raw data. It's very complete, but difficult to absorb. How can we summarize it in a way that gives a true picture of how those who work at the medical center are compensated?

One way would be to take the **arithmetic mean**—the average—of the data. To get the arithmetic mean we divide the total, $2,922,893, by the **tally**, the number of data points, which in this case is 57. $2,922,893 / 57 = $51,279

We could then say, "The **mean** annual salary at the medical center is $51,279." But would that give a true picture of what workers at the medical center are earning? No. If you look at the data, you can see that 11 employees make much more than $51,279, most of them more than twice as much, and everybody else makes less, most of them a great deal less. Clearly, the mean does not give a true picture of this data.

*(continued)*

Another way to summarize data is to find the **median.** The median is the number that is in the middle of the data. Half of the data points are above the median and half are below it. Since we have 57 data points, an odd number, we can find the median by counting down from the top (or up from the bottom) to the 29th data point. The median of this data is $40,544. (If we had an even tally, we would have to find the two points that are closest to the middle and take the mean of those two points. The number we got would be the median, even though it is not necessarily in the data set.)

If we say, "The median salary at the medical center is $40,544," does that give a truer picture of what workers at the medical center are earning? Perhaps. Notice that the median is more than $10,000 less than the mean.

An even better way to summarize this data would be to find the **quartiles.** Just as the median divides the data set into a top half and a bottom half, the quartiles divide each of the halves in half. The upper quartile is the median of the top half of the data set. Because there are 28 data points, an even number, in the top half of the data set, above the median, we'll take the mean of the two data points closest to the middle of the top half, the 14th and 15th, $48,444 and $47,614. ($48,444 + $47,614)/2 = $48,029. So the upper quartile is $48,029. The lower quartile, the median of the bottom half of the data set, calculated the same way, is $27,016.

Now we have three numbers, the upper quartile, $48,029, the median, $40,544, and the lower quartile, $27,016. We're getting a better picture of the data, but let's make it even better by adding two more numbers: the **range** and the interquartile range. The range is the difference between the **maximum** and the **minimum**—the largest data point minus the smallest. $254,949 – $15,879 = $239,070. The interquartile range is the difference between the upper quartile and the lower quartile. $48,029 – $27,016 = $21,013. Half of the data points are inside the interquartile range.

Now we have five numbers that, taken together, give a pretty good picture of the medical center payroll. The trouble is, most people can't keep five different numbers in their minds at the same time. So to present this kind of data visually, statisticians often use a **box plot.** A box plot is a number line with data plotted on it like this.

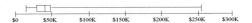

The single horizontal line above the number line goes from the minimum to the maximum. Its length is the range. The box begins at the lower quartile and ends at the upper quartile. Its length is the interquartile range. The vertical line inside the box is the median. The three vertical lines that make up the box divide the tally into four quarters. Half of the data points are in the box, a quarter of them below the box, and a quarter above. The box plot clearly shows that even though there is a wide range of salaries between the lowest-paid aide and the CEO, the majority of people who work at the medical center make between $27,000 and $48,000.

# Visualizing Data *(continued)*

**Activity 2**    **Developing Vocabulary in Context**

Read each sentence below. Choose the word or phrase from the box that best completes each sentence.

| | | |
|---|---|---|
| statistics | median | maximum |
| arithmetic mean | quartiles | minimum |
| tally | range | box plot |
| mean | | |

1. A graph that plots the maximum, minimum, quartiles, and median against a number line is a(n) _____.

2. Although there is such a thing as a geometric mean, when statisticians say _____ they usually mean _____.

3. Half of the data points are larger than the _____ and half are smaller.

4. The _____ and the median divide the data into four quarters.

5. The branch of math that is about collecting, interpreting, and presenting data is called _____.

6. The difference between the maximum and the minimum is the _____.

7. The greatest data point is called the _____.

8. The number of data points in a data set is called the _____.

9. The smallest data point is called the _____.

*Content-Area Vocabulary Strategies: Mathematics*

# Visualizing Data *(continued)*

**Activity 3**     **Extending Vocabulary Strategies**

Answer the following questions based on this data set. Round off your answers to the nearest whole number: 26, 11, 60, 91, 58, 88, 79, 9, 56, 60.

1. What is the arithmetic mean of the data?

2. What is the interquartile range?

3. What is the lower quartile?

4. What is the maximum?

5. What is the median?

6. What is the minimum?

7. What is the range?

8. What is the tally?

9. What is the upper quartile?

10. In the space below, draw a box plot of the data.

# Visualizing Data *(continued)*

**Activity 4**    **Writing to Build Vocabulary**

Create a data set of your own, real or imaginary. Analyze it and present it in a box plot. Write a brief analysis of your data. Use as many of the following words as you can in your analysis: *arithmetic mean, lower quartile, maximum, mean, median, minimum, upper quartile, range, statistics, tally.*

**Word Journal**

Are there words from this lesson, or other words you have found on your own, that you'd like to know more about? In a word journal, add any interesting words that you want to "make your own." Write the word and its definition. Add examples or drawings that explain the word.

# LESSON 6
## Measuring Angles

**Activity 1**     **Introducing Vocabulary in Context**

Read the following article. Notice the words in bold type. Use context clues to figure out the meaning of these words.

You have probably used a **protractor** to measure an **angle.** The system we use for measuring angles **derives** from the fundamental ideas of geometry. It's important to remember that geometry is about ideas, not drawings. We make drawings in order to communicate geometric ideas, but these are only approximations of the ideas we are communicating.

One of the fundamental ideas of geometry is a **point.** A point is something that we think of as so small that it has a location but no dimensions. For example, the sharp end of a needle, a dot that shows where Chicago is on a map of the world, or a star seen with the naked eye. In geometric drawings on paper or on a computer screen, we often use a dot to represent a point. We usually label points with a letter, like this.

. *A*

Another fundamental idea of geometry is a **line.** In geometry when we say line we always mean a line that is perfectly straight, like a beam of light, and extends forever in two opposite directions. When we draw a line we indicate that with arrowheads.

A line is a set of points. Because a

point has no dimensions, every point is infinitely small. That means every line is an infinite set of points. Think of any two points on a line. No matter how close they are to each other, there are always an infinite number of points in between them.

When two lines cross, there is always exactly one point—no more, no less— that belongs to both lines. That point is called the intersection of the two lines. Point A is the **intersection** of these two lines.

How many other lines could intersect at point *A*? An infinite number. Because each line is a set of points that are infinitely small, no matter how close to each other two lines that cross at *A* are, there can always be an infinite number of lines in between them. We can only draw a few of them. You have to imagine the rest.

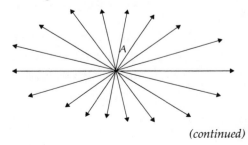

*(continued)*

         *Content-Area Vocabulary Strategies: Mathematics*

If there are two points, *A* and *B*, how many lines are there that contain both *A* and *B*? Exactly one line—no more, no less. So if we want to talk about a particular line, we can identify it by naming two points that are on it. For example, we can identify the line below by saying "line *AB*" (or "line *BA*.") That's just a short way of saying, "The line with points *A* and *B* on it."

A     B

Another fundamental idea of geometry is a **plane.** A plane is a flat surface with no edges. It goes on and on forever in two dimensions. Our world is full of things that look like planes or parts of planes: walls, floors, tabletops, window glass, a page of a book, the surface of a body of water. Under a microscope, they might look bumpy. For example, we know that the surface of the ocean is often wavy. But the idea of a plane is a set of points that is perfectly flat, infinitely long and wide, and one point thick.

When two planes cross, there is exactly one line—no more, no less— that belongs to both of them. That line is the intersection of the two planes. For example, look at the line where the floor of the room you're in meets a wall or the line where a wall meets the ceiling.

If two planes intersect at a line, how many other planes could intersect these two planes at the same line? An infinite number. Each plane is a set of points that are infinitely small. So no matter how close to each other two planes that cross at the line are, there can always be an infinite number of planes in between them. The way the pages of a book meet at the binding

suggests the way many planes can intersect in one line.

If there are three points, *A*, *B*, and *C* that are not all on the same line, how many planes are there that contain all three points?

.*A*

.*B*     .*C*

Clearly, the plane of this page contains all three points. Could there be another plane that contains them all? No. It's easy to visualize many planes that contain two of the three points, but there is only one that includes all three. For any three points that are not on the same line, there is always exactly one plane—no more, no less— that contains them all. If we want to talk about a particular plane, we can identify it by naming three points that are on it. For example, we can identify the plane of this page by saying "plane *ABC*" (or "plane *BCA*" or "plane *CAB*," and so on). That's just a short way of saying, "The plane with points *A*, *B*, and *C* on it."

We often want to talk about various parts of a line instead of all of it. The part of a line that goes from one point to another is called a **line segment.** A line segment just means two points and all the points between them on the line that goes through them. "Line segment *AB*" or "segment *AB*" are short ways of saying "points *A* and *B* and all the points between them on the line that contains *A* and *B*." "Segment *BA*" means the same thing. Points *A* and *B* are called the **endpoints** of segment *AB*.

*(continued)*

Another part of a line we need to be able to talk about is a ray. A **ray** is a part of a line that starts with a point on the line and goes on endlessly in one direction. "Ray *AB*" is a short way of saying, "Line segment *AB* and all the points on line *AB* that have *B* between them and *A*."

The point where the ray starts is called the endpoint. When we talk about a ray, we always give the endpoint first. The ray above is ray *AB*. Ray *BA* looks like this.

Two rays that have the same endpoint form an angle. For example,

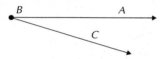

We always identify an angle by naming a point on one ray, then the endpoint, then a point on the other ray, in that order. The angle above is angle *ABC* or angle *CBA*. The letter in the middle tells you which point is the endpoint of the two rays.

Each of the rays that make up an angle is called a side. The common endpoint of the two rays is called the **vertex.** The sides of angle *ABC* are ray *BA* and ray *BC*. The vertex of angle *ABC* is point *B*.

Now that we've reviewed some of the fundamental ideas of geometry, we come to the subject at hand, measuring angles.

When the two rays that make up an angle belong to the same line, we call it

a straight angle. Angle *DEF* is a **straight angle.**

We know that there are an infinite number of other lines that could be drawn that intersect line *DF* at *E*. So we'll assign numbers to all the rays that all those lines create. We'll assign the number 0 to ray *ED* and the number 180 to ray *EF*. Then we'll assign all the **real numbers** between 0 and 180 to all the rays between ray *ED* and ray *EF* above line *DF*.

Mathematicians call this mapping. We are mapping the infinite set of real numbers from 0 to 180 onto the infinite set of rays from ray *ED* to ray *EF* above line *DF*. That means that for every real number there is a ray, and for every ray there is a real number.

Here are some of the rays and the numbers mapped onto them.

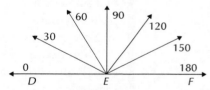

Using this system, we can measure angles by comparing them to the angles we have assigned numbers to. That's what you are doing when you measure an angle using a protractor. The unit of measure is called a **degree.** The symbol for degrees is °. If you write 180°, it's the same as writing "180 degrees."

Now that we have a measuring system, we can measure angles and classify them according to their measures. There are four classes: straight, right, acute, and obtuse.

*(continued)*

*Content-Area Vocabulary Strategies: Mathematics*

# Measuring Angles *(continued)*

A straight angle has a measure of 180°. An angle whose measure is 90° is a **right angle.** Angle *ABC* and angle *CBD* are right angles.

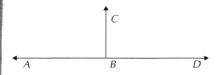

An angle whose measure is between 0° and 90° is an **acute angle.** Angle *FEG*, angle *FEH*, and angle *GEH* are acute angles.

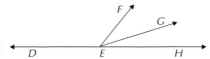

An angle whose measure is between 90° and 180° is an **obtuse angle.** Angle *DEF*, and angle *DEG* are obtuse angles.

Like points, lines, and planes, angles are geometric ideas. But they have real uses. That is why people have been measuring angles for thousands of years. Early astronomers tracked the apparent movement of the stars and planets to determine the best time to plant crops. They measured horizontal angles between celestial bodies and the four directions of the compass and vertical angles up from the horizon. From ancient times, sailors have navigated by measuring the angle between the direction of the boat and objects on the shore, as well as by measuring the angles to the stars and planets.

# Measuring Angles *(continued)*

**Activity 2**     **Developing Vocabulary in Context**

Read each sentence below. Choose the word or phrase from the list that best completes each sentence.

| | | |
|---|---|---|
| protractor | plane | real numbers |
| angle | line segment | degree |
| derive | endpoint | right angle |
| point | ray | acute angle |
| line | vertex | obtuse angle |
| intersection | straight angle | |

1. A(n) _____ is a device for measuring angles.

2. An angle is measured in a unit called a(n) _____ .

3. A set of _____s that is perfectly straight and extends indefinitely in two opposite directions is a(n) _____.

4. A(n) _____ is an angle of 90 _____s.

5. An angle of 100° is a(n) _____.

6. An angle of 180° is a(n) _____.

7. An angle of 50° is a(n) _____.

8. Most of geometry _____s from a few fundamental ideas.

9. Points R and S and all the points between them on line RS are a(n) _____s.

10. The set of _____ is an infinite set of numbers, including those that can be expressed in fractions as well as those that cannot.

11. The _____ of two lines is a(n) _____.

12. The common endpoint of the two _____s that make an angle is called the _____.

13. The geometric idea suggested by the surface of a sports playing field is a(n) _____.

14. Two _____s that have the same _____ make a(n) _____.

# Measuring Angles *(continued)*

**Activity 3**     **Extending Vocabulary Strategies**

Use a dictionary to help you to complete the following.

1. Give a brief definition of the word *degree* that covers its meaning when it is used outside of mathematics. Then write a sentence using that meaning.

   definition: _____

   sentence: _____

2. Give a brief definition of the word *obtuse* that covers its meaning when it is used outside of mathematics. Then write a sentence using that meaning.

   definition: _____

   sentence: _____

3. The word *acute* comes from the Latin word for needle. Find another word that comes from the same root and give a brief definition of it.

   word: _____

   definition: _____

4. The word *intersect* comes from the Latin words for between and cut. Find another word from the same root that begins with sect- and give a brief definition of it.

   word: _____

   definition: _____

5. What is the origin of the word *derive*? _____

   _____

**Activity 4**     **Writing to Build Vocabulary**

We are solid figures who live in a three-dimensional space surrounded by other solid figures. Imagine what it would be like to be a plane figure living in a two-dimensional space. Write a letter to someone who lives in three-dimensional space explaining what it is like to live in a plane. Use as many of the following words as you can: *acute angle, angle, degree, derive, endpoint, intersection, line, line segment, obtuse angle, plane, point, protractor, ray, real numbers, right angle, straight angle, vertex (vertices).*

**Word Journal**

Are there words from this lesson, or other words you have found on your own, that you'd like to know more about? In a word journal, add any interesting words that you want to "make your own." Write the word and its definition. Add examples or drawings that explain the word.

# L ESSON 7
## Angle Relationships

**Activity 1**      **Introducing Vocabulary in Context**

Read the article below. Notice the words in bold type. Use context clues to figure out the meaning of these words.

Some things about angles are always true. Artists, designers, engineers, surveyors, and others who use geometry in their work have found them useful over hundreds of years.

If there are two lines, they must relate to each other in one of three ways. They can be **skew,** they can intersect, or they can be **parallel.**

Skew lines are lines that are not in the same plane. Think of two jet planes and their condensation trails flying at different altitudes and in different directions.

Alternatively, if two lines are in the same plane, they either intersect or they don't. If they don't intersect, then they are parallel. Here are two parallel lines in the plane of this page.

When two lines intersect, they form angles. There are six angles in the drawing below: two straight angles, *ABC* and *DBE*; two acute angles, *ABD* and *EBC*; and two obtuse angles, *ABE* and *DBC.*

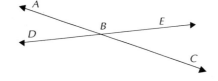

You can also see in the drawing that each of these pairs of angles is equal to each other. Angles *ABC* and *DBE* are equal to each other because they are both straight angles. The two acute angles and the two obtuse angles are formed by the same lines. The rays that form one are the extensions of the rays that form the other. Angles that are related to each other in this way are called **vertical angles.** Vertical angles are always equal.

Now let's look at angles *ABD* and *ABE*. One ray, *BA,* belongs to both of them. The other rays, *BD* and *BE* are parts of the same line. Together they make a straight angle, *DBE.* Angles that are related to each other in this way are called **supplementary angles.** Although we don't know exactly what their measures are, we know that the sum of their measures is 180°. There are four pairs of supplementary angles in the drawing: *ABD* and *ABE*, *ABE* and *EBC*, *EBC* and *CBD*, *CBD* and *DBA*.

Now we're going to add another line to the drawing: line *FG* that intersects line *AC* at *C.* Line *FG* is parallel to *DE.* A line like *AC* that cuts across two other lines is called a **transversal.** We now have two more acute angles and two more obtuse angles. All these

*(continued)*

angles are related to each other in special ways.

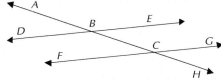

Because lines *DE* and *FG* are parallel, all the acute angles and all the obtuse angles are equal to each other. Let's look at the acute angles first. Angles *CBE* and *FCB* are both inside the parallel lines and on opposite sides of the transversal. They are called **alternate interior angles.** Alternate interior angles with a transversal across parallel lines are always equal.

Angles *ABD* and *CGH* are both outside the parallel lines and on opposite sides of the transversal. They are **alternate exterior angles.** Alternate exterior angles with a transversal across parallel lines are always equal.

Angles *ABD* and *BCF* are formed by the same transversal and by parallel lines. They are **corresponding angles** on the same side of a transversal across parallel lines. Corresponding angles on the same side of a transversal across parallel lines are always equal.

And we already know that angles *BCF* and *CGH*, like angles *ABD* and *EBC*, are vertical angles and therefore equal to each other. So all four acute angles in the drawing are equal. In the same way, all four obtuse angles are equal.

Now we're going to draw another line *IJ* through *B* that forms right angles with *DE* and *FG*. Because it forms right angles, it is called a

**perpendicular.** *IJ* is perpendicular to *DE* and to *FG*. It also forms some new angles, for example: *IBA*.

We still don't know exactly what the measure of angle *ABD* is, but we can see that *ABD* and *IBA* have a ray in common, *BA*, and their other rays, *BI* and *BD* form a right angle. Angles that are related to each other in this way are called **complementary angles.** We don't know what the measure of angle *ABD* or angle *IBA* is, but we know that the sum of their measures is 90°.

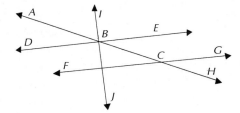

When you study the geometry of triangles and other polygons, you'll find that knowing the vocabulary of angles is useful. The words that students usually have the most trouble with are complementary and supplementary, probably because they are easily confused with each other. Not only that, complementary is easily confused with complimentary, which is a different word. Here are some mnemonic devices that may be helpful: 90° comes before 180°, *c* comes before *s* in the alphabet; supplementary angles add up to a straight angle; supplementary and straight both begin with *s*; something that is complementary completes something; something complimentary is something flattering or something free, which is what I like to get.

# Angle Relationships *(continued)*

**Activity 2**     **Developing Vocabulary in Context**

Read each sentence below and refer to the drawing. Choose the word or phrase from the box that best completes each sentence. Angles 7 and 10 are right angles.

| | |
|---|---|
| skew | alternate interior angles |
| parallel | alternate exterior angles |
| vertical angles | corresponding angles |
| supplementary angles | perpendicular |
| transversal | complementary angles |

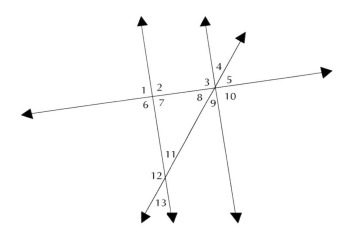

1. A line that forms a right angle with another line is called

   _____.

2. Alternate interior angles with a transversal are equal only if the lines crossed are _____.

3. Angles 11 and 12 are _____.

4. Angles 4 and 11 are _____ on the same side of a _____.

5. Angles 4 and 13 are _____.

6. Angles 4 and 9 are _____.

7. Angles 8 and 9 are _____.

8. Angles 9 and 11 are _____.

9. Two lines that are not in the same plane are _____ lines.

# Angle Relationships *(continued)*

**Activity 3**    **Extending Vocabulary Strategies**

Use a dictionary to help you to complete the following.

1. Give a brief definition of each word as it is used outside of mathematics. Then write a sentence for each using that meaning.

   a. alternate

   b. exterior

   c. interior

   d. vertical

   e. complementary

   f. corresponding

   g. skew

   h. supplementary

2. The word *perpendicular* comes from the Latin words *per*, which means through and *pend* or *pens*, which means to hang or weigh. Find a word that derives from either of these Latin words and give a brief definition. *Pend, pens* words may begin with *ap-, com-, de-, ex-, im-, pro-,* or *sus-*.

   word: _____

   definition: _____

3. The words *transversal, vertex,* and *vertical* come from the Latin word for turn. Find a word that derives from the same root; give a brief definition. They may begin with *ad-, con-, di-, extro-, intro-, in-,* or *uni-*.

   word: _____

   definition: _____

# Angle Relationships *(continued)*

**Activity 4**    **Writing to Build Vocabulary**

Choose one of the following ideas to write about.

- When computing work is done by many small computers each working on a small part of a problem instead of one big computer, it is called parallel processing. When two small children play next to each other, each with his or her own toys, talking out loud but not to each other, it is called parallel play. Describe an activity that you and others sometimes do that could be characterized as parallel and explain why the word *parallel* applies.

- Outside of mathematics, the word *corresponding* means agreeing or communicating by exchanging letters. Explain how those meanings relate to the meaning of *corresponding* in geometry.

## Word Journal

Are there words from this lesson, or other words you have found on your own, that you'd like to know more about? In a word journal, add any interesting words that you want to "make your own." Write the word and its definition. Add examples or drawings that explain the word.

# LESSON 8
## Platonic Solids

**Activity 1**    **Introducing Vocabulary in Context**

Read the following article. Notice the words in bold type. Use context clues to help you figure out the meaning of these words.

Much of the mathematics we study today was developed by the intellectuals of ancient Greece. Many of them believed that ideas, things that we perceive with our minds, are more real than things we perceive with our senses. The Greek idealists loved ideas of all kinds like beauty, justice, and virtue. They also loved mathematical ideas like geometric figures, especially **regular polygons, polygons** that have all sides and all angles equal. For example,

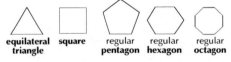

equilateral triangle   square   regular pentagon   regular hexagon   regular octagon

The Greeks were passionate about what we now call Platonic solids, after Plato, the leading idealist philosopher. A solid figure is a platonic solid if

1. all vertices are exactly alike.
2. all **faces** are regular polygons and exactly alike.

How many Platonic solids are there? It turns out there are five of them.

Since all the vertices and all the faces have to be exactly alike, we can define a Platonic solid by just two numbers:

1. the number of faces that meet at one vertex.
2. the number of sides each face has.

Clearly, you can't have a solid figure with zero, one, or two sides or faces. A figure that doesn't have at least three faces meeting at a vertex and three sides on each face won't be solid.

If each face has three sides, that means that each face is an **equilateral triangle.** If three equilateral triangles meet at each vertex, does that make a Platonic solid? Yes it does. It makes what is called a tetrahedron, which is Greek for four faces. A tetrahedron has four vertices and looks like this:

What if each face has three sides and *four* faces meet at each vertex? That works, too. What you get is an octahedron (eight faces), a figure with six vertices. It looks like this:

What if each face has three sides and *five* faces meet at each vertex? That's called an icosahedron. It has 20 faces and 12 vertices. It looks like this:

*(continued)*

*Content-Area Vocabulary Strategies: Mathematics*

What if each face has three sides and *six* faces meet at each vertex? If you put six equilateral triangles together, you get a flat **hexagon,** not a solid figure. That's as far as we can go with three-sided faces.

What if each face has *four* sides and *three* faces meet at each vertex? That means each side is a **square.** You get a **cube,** a solid figure we have all known since we first played with blocks. It has six faces and six vertices. You know what that looks like.

What if each face has four sides and *four* faces meet at each vertex? If you put four squares together, you get a flat square, not a solid figure. That's as far as we can go with four-sided faces.

What if each face has *five* sides and *three* faces meet at each vertex? You get

a dodecahedron (12 faces). It has 12 vertices. It looks like this.

What if each face has *six* sides and *three* faces meet at each vertex? If you put three regular hexagons together, you get a flat figure that looks like this, not a Platonic solid.

So the ancient Greeks concluded that there are five Platonic solids, no more, no less, each of them perfectly symmetrical with every face, every vertex, every edge, and every angle exactly identical. Can such perfection exist in our world of everyday reality? Or only in the realm of pure thought? The Greeks had their answer. What's yours?

     *Content-Area Vocabulary Strategies: Mathematics*

# Platonic Solids (continued)

**Activity 2**      **Developing Vocabulary in Context**

Read each sentence below. Choose the word or phrase from the box that best completes each sentence.

| | | |
|---|---|---|
| regular polygons | pentagon | solid |
| polygons | hexagon | faces |
| equilateral triangle | octagon | cube |
| square | | |

1. Closed figures made up of line segments are called _____.

2. A geometric figure that exists in three dimensions is a(n) _____ figure.

3. Plane surfaces that bound solid figures are _____.

4. A rectangle with all four sides the same length is a(n) _____.

5. A polygon with eight sides is a(n) _____.

6. Polygons with equal sides and equal angles are _____.

7. A polygon with five sides is a(n) _____.

8. A polygon with six sides is a(n) _____.

9. A regular polygon with three sides is a(n) _____.

10. A solid figure with six equal square faces is a(n) _____.

# Platonic Solids *(continued)*

**Activity 3**    **Extending Vocabulary Strategies**

Draw each of the following.

1. a hexagon

2. an octagon

3. a pentagon

4. a polygon that is not regular

5. Draw *one* of the following: a tetrahedron, an octahedron, an icosahedron, a cube, or a dodecahedron. On your drawing, label a vertex and a face.

**Activity 4**  **Writing to Build Vocabulary**

The ancient Greek idealists, notably Plato, believed that ideas, things like Platonic solids that we perceive with our minds, are more real than things we perceive with our senses. Ideas are perfect, beautiful, and eternal, they believed; the things we have around us in our lives, man-made and natural—houses, trees, dogs, chairs, even people—are imperfect and temporary, poor imitations of the real thing in our minds. Do you agree? Why or why not? Use some of the following words in your answer: *cube, dodecahedron, equilateral triangle, face, hexagon, icosahedron, octagon, octahedron, pentagon, polygon, regular polygon, solid, tetrahedron.*

**Word Journal**

Are there words from this lesson, or other words you have found on your own, that you'd like to know more about? In a word journal, add any interesting words that you want to "make your own." Write the word and its definition. Add examples or drawings that explain the word.

# QUIZ: Lessons 5–8

To answer questions 1—5, refer to the following data set:

3, 4, 5, 10, 10, 10, 10, 10, 20

1. The arithmetic mean is
   (a) 4.5.
   (b) 9.1.
   (c) 10.
   (d) 17.
   (e) 20.

2. The maximum is
   (a) 4.5.
   (b) 9.1.
   (c) 10.
   (d) 17.
   (e) 20.

3. The lower quartile is
   (a) 4.5.
   (b) 9.1.
   (c) 10.
   (d) 17.
   (e) 20.

4. The median is
   (a) 4.5.
   (b) 9.1.
   (c) 10.
   (d) 17.
   (e) 20.

5. The range is
   (a) 4.5.
   (b) 9.1.
   (c) 10.
   (d) 17.
   (e) 20.

To answer questions 6–10, refer to this drawing. Angle *DEC* is a right angle.

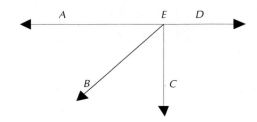

6. Angle *AEB* is a(n) _____ angle.

    (a) acute

    (b) obtuse

    (c) right

    (d) straight

7. Angle *BED* is a(n) _____ angle.

    (a) acute

    (b) obtuse

    (c) right

    (d) straight

8. Angle *AEC* is a(n) _____ angle.

    (a) acute

    (b) obtuse

    (c) right

    (d) straight

9. Angle *AED* is a(n) _____ angle.

    (a) acute

    (b) obtuse

    (c) right

    (d) straight

10. Point *E* is the _____ of angle *BEC*.

    (a) endpoint

    (b) intersection

    (c) vertex

    (d) all of the above

*Content-Area Vocabulary Strategies: Mathematics*

To answer questions 11–15, refer to this drawing.

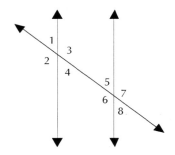

11. Which angles are vertical angles?
    (a) 2 and 7
    (b) 3 and 4
    (c) 3 and 6
    (d) 6 and 7

12. Which angles are supplementary angles?
    (a) 2 and 7
    (b) 3 and 4
    (c) 3 and 6
    (d) 6 and 7

13. Which angles are alternate interior angles?
    (a) 2 and 7
    (b) 3 and 4
    (c) 3 and 6
    (d) 6 and 7

14. Which angles are alternate exterior angles?
    (a) 2 and 7
    (b) 3 and 4
    (c) 3 and 6
    (d) 6 and 7

15. If the lines in the drawing that appear to be parallel really are parallel, which pairs of angles are not equal?
    (a) 2 and 7
    (b) 3 and 4
    (c) 3 and 6
    (d) 6 and 7

# Platonic Solids *(continued)*

To answer questions 16–20, refer to these figures by number.

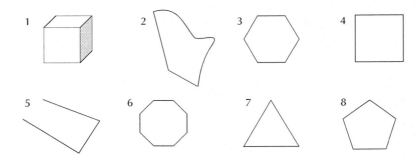

16. Which figure is a cube?

    (a) 1

    (b) 4

    (c) 5

    (d) none of the above

17. Which figure is a hexagon?

    (a) 3

    (b) 4

    (c) 6

    (d) 8

18. Which figure is an octagon?

    (a) 3

    (b) 4

    (c) 6

    (d) 8

19. Which figure is a pentagon?

    (a) 3

    (b) 4

    (c) 6

    (d) 8

20. Which figure is a polygon?

    (a) 1

    (b) 2

    (c) 5

    (d) 7

# GLOSSARY

**accuracy (accurate):** the closeness of a measure to a standard or true value (3)

**acute angle:** an angle whose measure is less than 90 degrees (6)

**alternate exterior angles:** angles formed by parallel lines and a transversal outside the parallel lines on opposite sides of the transversal (7)

**alternate interior angles:** angles formed by parallel lines and a transversal inside the parallel lines on opposite sides of the transversal (7)

**angle:** two rays with a common endpoint (6)

**arithmetic mean:** the total of the data divided by the tally; the average (5)

**attribute:** a characteristic or quality belonging to a specific person or thing (3)

**base (of an exponent):** a number that is raised to a power (1)

**binary system:** the base two system of numerals (1)

**box plot:** a graph that plots the minimum, the quartiles, the median, and the maximum against a number line (5)

**complementary angles:** angles that have a ray in common and whose other rays form a right angle (7)

**composite (number):** a counting number that is not 1 and not prime (2)

**coordinate:** a number used to specify the location of a point on a number line or a graph (4)

**corresponding angles:** angles formed by parallel lines and a transversal on the same side of the parallel lines and on the same side of the transversal (7)

**cube:** a solid figure with six equal square sides (8)

**customary (system):** the traditional system of units of measurement based on the human body: feet, inches, pounds, ounces, and so forth (3)

**decimal system:** the system of numerals based on ten (1)

**degree:** a unit of measure of an angle equal to 1/180th of a straight angle (6)

**derive:** to derive one idea from another is to go from the first to the second by logical reasoning (6)

**difference:** the result of subtracting numbers (4)

**digit:** one of the set of symbols used in the decimal system or a system with another base (1)

**divisible:** capable of being divided with no remainder (2)

**endpoint:** a point in a line segment or a ray that is not between two other points in the line segment or ray (6)

**equilateral triangle:** a triangle whose sides and angles are all equal (8)

**error (of measurement):** the difference between an observed value and a true value (3)

**evaluate:** to find the value of; in mathematics, to reduce to simplest form (1)

**expanded form:** to express a number as the sum of the values represented by each individual digit; for example, 123 in expanded form is 100+20+3 (1)

**exponent:** a symbol written above and to the right of a mathematical expression to indicate the operation of raising to a power (1)

**faces:** plane surfaces that bound solid figures (8)

**factor:** any one of a set of numbers to be multiplied (2)

# Glossary (continued)

**gram:** a metric unit of mass equal to 1/1000 kilogram and nearly equal to the mass of one cubic centimeter of water at its maximum density (3)

**greatest possible error:** one half of the unit used in a measurement (3)

**hexagon:** a polygon with six sides (8)

**identity property of multiplication:** the product of any number and 1 is the number itself: $a \times 1 = a$ (2)

**intersection:** in geometry, the set of points that belong to both one set and the other (6)

**line:** a set of points that is straight, like a beam of light, and extends indefinitely in two opposite directions (6)

**line segment:** the set of two points on a line and all points between them (6)

**maximum:** the greatest data point (5)

**mean:** the arithmetic mean unless otherwise stated (5)

**measure:** to assign a number to a thing's attributes, such as height, weight, and so forth (3)

**median:** the number that is less than half of the data points and greater than the other half of them (5)

**meter:** the base unit of length in the metric system, equal to the distance traveled by light in a vacuum in 1/299,792,458 second or to about 39.37 inches (3)

**metric (system):** a decimal system of weights and measures based on the meter and on the kilogram (3)

**minimum:** the least data point (5)

**multiple:** the product of a number by a counting number (2)

**negative slope:** a slope between two points on a graph where $y$ decreases as $x$ increases (4)

**obtuse angle:** an angle whose measure is between 90 and 180 degrees (6)

**octagon:** a polygon with eight sides (8)

**ordered pair:** a pair of numbers that go together in a certain order (4)

**origin:** the (0, 0) point of a graph (4)

**parallel:** lines in the same plane that do not intersect (7)

**pentagon:** a polygon with five sides (8)

**perpendicular:** forming a right angle (7)

**place value:** the value assigned to a certain position in a system of numerals (1)

**plane:** a flat surface that extends indefinitely in two dimensions (6)

**point:** something so small it has a location but no dimensions (6)

**polygons:** closed plane figures made up of line segments (8)

**positive slope:** a slope between two points on a graph where $y$ increases as $x$ increases (4)

**power:** the number of times as indicated by an exponent that a number occurs as a factor in a product; also, the product itself (1)

**precision (precise):** the degree of refinement with which an operation is performed or a measurement stated (3)

**prime (number):** a counting number that has exactly two factors: itself and 1 (2)

**product:** the result of multiplication (2)

**protractor:** a device for measuring angles (6)

**quartiles:** the median of the half of the data below the median and the half above the median (5)

**quotient:** the result of division (2)

**random:** a set of events each of which is equally likely to occur or a process that obtains them; unpredictable (1)

**range:** the difference between the maximum and the minimum (5)

**ray:** ray $AB$ is line segment $AB$ and all points on line $AB$ that have $B$ between it and $A$ (6)

**real numbers:** the set of all rational and irrational numbers (6)

**regular polygons:** polygons with sides of equal length and angles of equal measure (8)

**right angle:** an angle whose sides are perpendicular; a 90° angle (6)

# Glossary *(continued)*

**Sieve of Eratosthenes:** an easy method of finding prime numbers; named after Eratosthenes, a Greek philosopher (2)

**skew (lines):** lines not in the same plane (7)

**solid:** geometric figure that exists in three dimensions (8)

**square:** a rectangle with all four sides equal (8)

**square root:** a factor of a number that when squared (multiplied by itself) gives the number itself; for example, the square root of 9 is ± 3 (2)

**statistics:** the branch of mathematics concerned with collecting, analyzing, interpreting, and presenting data (5)

**straight angle:** an angle whose rays belong to the same line (6)

**supplementary angles:** angles that have a ray in common and whose other rays form a straight angle (7)

**tally:** the number of data points in a data set (5)

**transversal:** a line that intersects two other lines in the same plane (7)

**vertex (vertices):** the endpoint of the rays of an angle or the edges of a solid figure (6)

**vertical angles:** angles on opposite sides of intersecting lines (7)

***x*-axis:** the horizontal axis of a graph where the values of $x$ are represented (4)

***x*-coordinate:** the first number in an ordered pair (4)

***y*-axis:** the vertical axis of a graph where the values of $y$ are represented (4)

***y*-coordinate:** the second number in an ordered pair (4)

# ANSWER KEY

In Lessons 1–8, Activity 1 involves reading an article and has no keyed answers. Activity 4 is always a writing activity, and students' writings will vary.

## Introductory Lesson: Context Clues

1. a spending plan; definition: a budget is a spending plan, also, explanation: a budget can help you make wise decisions about your money
2. add up; restatement
3. all the money you take in; paycheck and allowance
4. things you spend money on; examples: monthly haircut, weekly DVD rental
5. figure out through mathematical operations; relates to expenses, which you add up, then deduct from income (appears later in the paragraph)
6. by a large amount, significantly, substantially; an antonym-type clue: if prices do not increase dramatically, they can be used for a reasonable calculation; therefore, opposite of *dramatically* must be *not much*
7. occasional extras; examples: birthday gift and meal out, and antonym-type clue: incidentals are not like regular expenses, which are fixed
8. do not vary; explanation: incidentals are not fixed, and incidentals vary, so something that is fixed does not vary
9. subtract; restatement
10. to spend on something unusual; description-type of clue: splurge is what you do when you buy incidentals

## Lesson 1: Counting in an On/Off Universe

### Activity 2

1. binary system, digit, evaluate, decimal system
2. expanded form, place value
3. random
4. base, exponent, power

### Activity 3

The answers to these questions will vary. You might have students work in groups to see which group can find the most words for each question.

## Lesson 2: Nonrectangular Numbers

### Activity 2

1. Sieve of Eratosthenes
2. composite
3. product, identity property of multiplication
4. prime
5. divisible, quotient, factor, multiple
6. square root

### Activity 3

1.–10. Answers will vary.

## Lesson 3: Real World Math

### Activity 2

1. gram, metric, customary
2. meter
3. precise, accurate
4. greatest possible error
5. attribute
6. error
7. measure

### Activity 3, Part I

1. precision
2. meter
3. attribute

**Part II**

1. I
2. C
3. O
4. I
5. M
6. C
7. M

## Lesson 4: Visualizing Measurements

**Activity 2**

1. ordered pair
2. negative slope
3. *x*-axis, *y*-axis
4. coordinate
5. positive slope
6. *x*-coordinate, *y*-coordinate
7. difference
8. origin

**Activity 3**

1. Positive slope
2. coordinate
3. ordered pair
4. *y*-axis
5. *x*-axis
6. *x*-coordinate, *y*-coordinate
7. negative slope
8. origin
9. d

## Quiz: Lessons 1–4

1. a
2. b
3. d
4. d
5. c
6. b
7. b
8. a
9. d
10. a
11. d
12. b
13. c
14. b
15. c
16. a
17. d
18. a
19. a
20. b

## Lesson 5: Visualizing Data

**Activity 2**

1. box plot
2. mean, arithmetic mean
3. median
4. quartiles
5. statistics
6. range
7. maximum
8. tally
9. minimum

**Activity 3**

1. 53.8
2. 53
3. 26
4. 91
5. 59
6. 9
7. 82
8. 10
9. 79
10.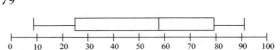

## Lesson 6: Measuring Angles

**Activity 2**

1. protractor
2. degree
3. point, line
4. right angle, degree
5. obtuse angle
6. straight angle
7. acute angle
8. derive
9. line segment
10. real numbers
11. intersection, point
12. rays, vertex
13. plane
14. ray, endpoint, angle

**Activity 3**

1. definition: a title conferred on students who complete a college program of study; Sentences will vary.
2. definition: dull; Sentences will vary.
3. Words and definitions will vary.
4. Words and definitions will vary.
5. derive—from the Latin *derivare*, to draw off

## Lesson 7: Angle Relationships

**Activity 2**

1. perpendicular
2. parallel
3. supplementary angles
4. corresponding angles, transversal
5. alternate exterior angles
6. vertical angles
7. complementary angles
8. alternate interior angles
9. skew

**Activity 3**

1.–3. Answers will vary.

## Lesson 8: Platonic Solids

**Activity 2**

1. polygons
2. solid
3. faces
4. square
5. octagon
6. regular polygons
7. pentagon
8. hexagon
9. equilateral triangle
10. cube

**Activity 3**

1. See drawings in the text.
2. See drawings in the text.
3. See drawings in the text.
4. See drawings in the text.
5. Answers will vary. See drawings in the text.

## Quiz: Lessons 5–8

| | | | |
|---|---|---|---|
| 1. b | | 11. d | |
| 2. e | | 12. b | |
| 3. a | | 13. c | |
| 4. c | | 14. a | |
| 5. d | | 15. b | |
| 6. a | | 16. a | |
| 7. b | | 17. a | |
| 8. c | | 18. c | |
| 9. d | | 19. d | |
| 10. c | | 20. d | |